Further Experiments and Observations in an Heated Room

BY

CHARLES BLAGDEN, M.D. F.R.S.

Further Experiments and Observations in an Heated Room

On the third of April, nearly the same party as before, together with Lord Seaforth, Sir George Home, Mr. Dundas, and Dr. Nooth, went to the heated room in which the experiments of the 23d of January were made. Dr. Fordyce had ordered the fire to be lighted the preceding day, and kept up all night; so that every thing contained in the room, and the walls themselves, being already well warmed, we were able to push the heat to a much higher degree than before. In the course of the day several different sets of experiments were going on together; but to avoid confusion, it will be necessary to relate each series by itself, without regard to the order of time; beginning with that series which serves as a continuation of our former experiments.

Soon after our arrival, a thermometer in the room rose above the boiling point; this heat we all bore perfectly well, and without any sensible alteration in the temperature of our bodies. Many repeated trials, in successively higher degrees of heat, gave still more remarkable proofs of our resisting power. The last of these experiments was made about eight o'clock in the evening, when the heat was at the greatest: a very large thermometer, placed at a distance from the door of the room, but nearer to the wall than to the cockle, and defended from the immediate action of the cockle by a piece of paper hung before it, rose one or two degrees above 260°: another thermometer, which had been suspended very near the door, stood some degrees above 240°. At this time I went into the room, with the addition to my common clothes, of a pair of thick worsted stockings drawn over my shoes, and reaching some way above my knees; I also put on a pair of gloves, and held a cloth constantly between my face and the cockle: all these precautions were necessary to guard against the scorching of the red-hot iron. I remained eight minutes in this situation, frequently walking about to all the different parts of the room, but standing still most of the time in the coolest spot, near the lowest thermometer. The air felt very hot, but still by no means to such a degree as to give pain: on the contrary, I had no doubt of being able to support a much greater heat; and all the gentlemen present, who went into the room, were of the same opinion. I sweated, but not very profusely. For seven minutes my breathing continued perfectly good; but after that time I began to feel an oppression in my lungs, attended with a sense of anxiety; which gradually increasing for the space of a minute, I thought it most prudent to put an end to the experiment, and immediately left the room. My pulse, counted as soon as I came into the cool air, for the uneasy feeling rendered me incapable of examining it in the room, was found to beat at the rate of 144 pulsations in a minute, which is more than double its ordinary quickness. To this circumstance the oppression on my breath must be partly imputed, the blood being forced into my lungs quicker than it could pass through them; and hence it may very reasonably be conjectured, that should an heat of this kind ever be pushed so far as to prove fatal, it will be found to have killed by an accumulation of blood in the lungs, or some other immediate effect of an accelerated circulation; for all the experiments show, that heating the air does not make it unfit for respiration, communicating to it no noxious quality except a power of irritating. In the course of this experiment, and others of the same kind by several of the gentlemen present, some circumstances occurred to us which had not been remarked before. The heat, as might have been expected, felt most intense when we were in motion; and, on the same principle a blast of the heated air from a pair of bellows was scarcely to be borne; the sensation in both these cases exactly resembled that felt in our nostrils on inspiration. The reason is obvious; when the same air remained for any time in contact with our bodies, part of its heat was destroyed, and consequently we came to be surrounded with a cooler medium than the common air of the room; whereas when fresh portions of the air were applied to our bodies in such a quick succession, that no part of it could remain in contact a sufficient time to be cooled, we necessarily felt the full heat communicated by the stove. It was observed that our breath did not feel cool to the fingers unless they were held very near the mouth; at a distance the cooling power of the breath did not sufficiently compensate the effect of putting the air in motion, especially when we breathed with force.

A chief object of this day's experiments was, to ascertain the real effect of our cloaths in enabling us to bear such high degrees of heat. With this view I took off my coat, waistcoat, and shirt, and in that situation went into the room, as soon as the thermometer had risen above the boiling point, with the precaution of holding a piece of cloth constantly between my body and the cockle, as the scorching was otherwise intolerable. The first impression of the heated air on my naked body was much more disagreeable than I had ever felt it through my cloaths; but in five or six minutes a profuse sweat broke out, which gave me instant relief, and took off all the extraordinary uneasiness: at the end of twelve minutes, when the thermometer had risen almost to 220°, I left the room, very much fatigued, but no otherwise disordered; my pulse made 136 beats in a minute. On this occasion I felt nothing of that oppression on my breath which became so material a symptom in the experiment with my clothes when the thermometer had risen to 260°: this may be partly explained by the less quickness of my pulse, the difference being at least eight beats in a minute, and probably more, as in the experiment without my shirt the pulsations were counted before I had left the room; but there is a further circumstance to be taken into consideration, that the experiment attended with oppression on the breath was made in the evening after a very plentiful meal, whereas the other was made in the forenoon, some hours after a moderate breakfast. The unusual degree of fatigue which I felt from the experiment without my shirt, must be ascribed in great measure to the more violent effort which the living powers were obliged to exert, in order to preserve the due human temperature, when such hot air came into immediate contact with my body. In the present case it appears beyond all doubt, that the living powers were very much assisted by the perspiration, that cooling evaporation which is a further provision of nature for enabling animals to support great heats. Had we been provided with a proper balance, it would undoubtedly have rendered the experiment more complete to have taken the exact weight of my body at going into, and coming out of, the room; as from the quantity lost, some estimate might be formed of the share which the perspiration had in keeping the body cool; probably its effect was very considerable, but by no means sufficient to account for the whole of the cooling, and certainly not equable enough to keep the temperature of the body to such an exact pitch: For it should here be remarked, that during all the experiments made this day, whenever I tried the heat of my body, the thermometer always came very nearly to the same point; I could not perceive even the small difference of one degree, which was observed in our former experiments. Should these considerations however be thought insufficient to prove that evaporation was not the sole agent in keeping the body cool, I believe that Dr. Fordyce's experiments in moist air will be found to remove all doubts on this subject. Several of the gentlemen present, as well as myself, went into the room without shirts many times afterwards, when the thermometer had risen much higher, almost to 260°, and found that we could bear the heat very well, though the first sensation was always more disagreeable than with our cloaths.

In all the experiments made this day it was observed, that the thermometer did not sink so much in consequence of our stay in the room as on the 23d of January; probably because a much larger mass of matter had been heated by the longer continuance of the fire.

Our own observations, together with those of M. Tillet, in the Memoirs of the Academy of Sciences, had given us good reason to suspect, that there must have been some fallacy in the experiment with a dog, made at the desire of Dr. Boerhaave, and related in his Elements of Chemistry.

To determine this matter more exactly, we subjected a bitch weighing thirty-two pounds, to the following experiment. When the thermometer had risen to 220°, the animal was shut up in the heated room, inclosed in a basket, that its feet might be defended from the scorching of the floor, and with a piece of paper before its head and breast to intercept the direct heat of the cockle. In about ten minutes it began to pant and hold out its tongue, which symptoms continued till the end of the experiment, without ever becoming more violent than they are usually observed in dogs after exercise in hot weather; and the animal was so little affected during the whole time, as to show signs of pleasure whenever we approached the basket. After the experiment had continued half an hour, when the thermometer had risen to 236°, we opened the basket, and found the bottom of it very wet with *saliva,* but could perceive no particular *foetor.* We then applied a thermometer between the thigh and flank of the animal; in about a minute the quicksilver sunk down to 110°: but the real heat of the body was certainly less than this, for we could neither keep the ball of the thermometer a sufficient time in proper contact, nor prevent the hair, which felt sensibly hotter than the bare skin, from touching every part of the instrument. I have since found, that the thermometer held in the same place, when the animal is perfectly cool and at rest, will not rise above 101°. At the end of thirty-two minutes the bitch was permitted to go out of the room; on coming into the cold air she appeared perfectly brisk and lively, not in the least injured by the heat, and has now continued very well above a month. Our experiment therefore differs, in every essential circumstance of the event, from that related by Dr. Boerhaave. With respect to this last it is remarkable, if the facts be properly represented, that an intolerable stench arose from the dog; and that an assistant dropped down senseless on going into the stove.

To prove that there was no fallacy in the degree of heat shown by the thermometer, but that the air which we breathed was capable of producing all the well-known effects of such a heat on inanimate matter, we put some eggs and a beef-steak on a tin frame, placed near the standard thermometer, and farther distant from the cockle than from the wall of the room. In about twenty minutes the eggs were taken out, roasted quite hard; and in forty-seven minutes the steak was not only dressed, but almost dry. Another beef-steak was rather over-done in thirty-three minutes. In the evening, when the heat was still greater, we laid a third beef-steak in the same place: and as it had now been observed, that the effect of the heated air was much increased by putting it in motion, we blew upon the steak with a pair of bellows, which produced a visible change on its surface, and seemed to hasten the dressing; the greatest part of it was found pretty well done in thirteen minutes.

About the middle of the day two similar earthen vessels, one containing pure water, and the other an equal quantity of the same water with a bit of wax, were put upon a piece of wood in the heated room. In one hour and an half the pure water was heated to 140° of the thermometer, while that with the wax had acquired an heat of 152°, part of the wax having melted and formed a film on the surface of the water, which prevented the evaporation. The pure water never came near the boiling point, but continued stationary above an hour at a much lower degree; a small quantity of oil was then dropped into it, as had before been done to that with the wax; in consequence of which, the water in both the vessels came at length to boil very briskly. A saturated solution of salt in water put into the room, was found to heat more quickly, and to an higher degree, than pure water, probably because it evaporated less; but it could not be brought to boil till oil was added, by means of which it came towards evening into brisk ebullition, and consequently had acquired an heat of 230°. Some rectified spirit of wine in a bottle slightly corked, which had been immersed into this solution of salt while cold, began to boil in about two hours, and soon afterwards was totally evaporated. Perhaps no experiments hitherto made furnish more remarkable instances of the cooling effect of evaporation than these last facts; a power which appears to be much greater than hath commonly been suspected. The evaporation itself, however, was more considerable in our experiments than it can be in almost any other situation, because the air applied to the evaporating surface was uncommonly hot, and at the same time not more charged with moisture than in its ordinary state. A powerful assistant evaporation must undoubtedly prove, in keeping the living body properly cool, when exposed to great heats; but it can act only in a *gross* way, and by no means in such a nice proportion to the momentary exigencies of the animal as would be requisite for the exact preservation of its temperature: that other provision of nature which seems more immediately connected with the powers of life, is probably the great agent in preserving the just balance of temperature; exerting a greater effort in proportion as the evaporation is deficient, and a less effort as the evaporation increases. This idea corresponds with the general analogy of the animal economy, the nicer balances of which are almost universally effected in that part of the body which is formed with the most subtle organization.

The heated room will, I hope, in time become a very useful instrument in the hands of the physician. Hitherto the necessary experiments have not been made to direct its application with a sufficient degree of certainty. However, we can already perceive a foundation for some distinctions in the use of this uncommon remedy. Should the object in view be to produce a profuse perspiration, a dry heat acting on the naked body would most effectually answer that purpose. The histories of dropsies and some other diseases, supposed to have been cured by such means, are well known to every physician. In some cases also, a moist heat, and in others heat transmitted through a quantity of cloaths, might have their peculiar advantages. That the danger likely to ensue from such applications is less than has been commonly apprehended, our former experiments gave sufficient reason to believe, and the same was amply confirmed by those which: make the subject of this paper. For during the whole day, we passed out of the heated room, after every experiment, immediately into the cold air, without any precaution; after exposing our naked bodies to the heat, and sweating most violently, we instantly went out into a cold room, and staid there even some minutes before we began to dress; yet no one received the least injury. I felt nothing this day of the noise and giddiness in my head, which had affected me in making the former experiments; and, whether from the force of habit, or any other cause, the shaking of our hands was less, and we felt less languor, though the beat had been so much more intense.

www.ingramcontent.com/pod-product-compliance
Lightning Source LLC
Chambersburg PA
CBHW070721210526
45170CB00021B/1394